baby SLOTHS

KIM THOMPSON

CREATIVE EDUCATION • CREATIVE PAPERBACKS

CONT

ENTS

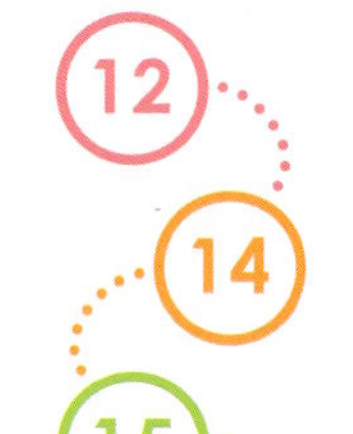

I AM A BABY SLOTH.

I am a baby three-toed sloth.
Count the claws on each foot: 1, 2, 3.

There are also two-toed sloths. You can see them at zoos.

My mom has just one baby at a time. I get a ride through the canopy of the rainforest.

When I feel brave, I let go of my mom. I reach out and grab branches.

I drink my mom's milk.

Then, I taste chewed-up leaves from her mouth. It helps me learn which leaves are good to eat.

I move slowly.
I sleep a lot of
the time.

After one year, I leave my mom. I am ready to find my own leafy home.

Green algae will grow on my fur.

SPEAK AND LISTEN

Can you speak like a baby sloth?

Baby sloths squeal and squeak.

Listen to these sounds:

https://www.youtube.com/watch?v=5aYnDI1MDk8

Now it is your turn!

SLOTH WORDS

algae: small plants without roots or stems that grow in wet places

canopy: a layer of the rainforest that is high up in tall trees

claws: sharp, hard nails on an animal's feet

rainforest: a thick forest in parts of the world that are always warm and that get lots of rain

READING CORNER

Kellett, Jenny. *The Ultimate Sloth Book for Kids*. Melbourne, Australia: Bellanova Books, 2022.

Leavitt, Amie Jane. *Awesome Animals of South America: the Continent and Its Creatures Great and Small*. Mount Joy, Penn.: Fox Chapel Publishing, 2024.

Rocco, Hayley. *Hello, I'm a Sloth (Meet the Wild Things)*. New York: G.P. Putnam's Sons Books for Young Readers, 2024.

INDEX

PUBLISHED BY CREATIVE EDUCATION AND CREATIVE PAPERBACKS
P.O. Box 227, Mankato, Minnesota 56002
Creative Education and Creative Paperbacks are imprints of The Creative Company
www.thecreativecompany.us

LIBRARY OF CONGRESS CATALOGING-IN-PUBLICATION DATA
Names: Thompson, Kim, 1970- author.
Title: Baby sloths / Kim Thompson.
Description: Mankato, Minnesota : Creative Education and Creative Paperbacks, [2026] | Series: Starting out | Includes bibliographical references and index. | Audience: Ages 4-7 | Audience: Grades K-1 |
Summary: "Introduce beginning readers to the world of baby sloths with this life science starter. Includes photos, a labeled animal diagram, "Make a Noise" section, glossary, and further resources"-- Provided by publisher.
Identifiers: LCCN 2024043263 (print) | LCCN 2024043264 (ebook) | ISBN 9798889897552 (library binding) | ISBN 9781682778418 (paperback) | ISBN 9798889897682 (ebook)
Subjects: LCSH: Sloths--Infancy--Juvenile literature.
Classification: LCC QL737.E2 T546 2026 (print) | LCC QL737.E2 (ebook) | DDC 599.3/131392--dc23/eng/20241220
LC record available at https://lccn.loc.gov/2024043263
LC ebook record available at https://lccn.loc.gov/2024043264

DESIGN AND PRODUCTION
Design by Rhea Magaro
Production by Beeline Media and Design, Inc.
Art direction by Tom Morgan

PHOTOGRAPHS by Alamy Stock Photo/Bos11, 14, imageBROKER/Moritz Wolf, 6-7, Michael S. Nolan, 9; Dreamstime/Artushfoto, 11, Isselee, 2, 4, 5, 12, Seadam, cover, 10-11; Shutterstock/Bohbeh, 13, J. Niens, 8, Mark_Kostich, 7

Printed in India